Fabulous Floor Patterns

S. F. "Jerry" Cook III
& Tina Skinner

Schiffer Publishing, Ltd.
4880 Lower Valley Rd.
Atglen, PA 19310 USA
www.schifferbooks.com

INTRODUCTION

Enjoy hundreds of floor patterns, from checkerboard patterns and wild tile configurations to floral borders and faux marble magic, drawn from advertising and promotional literature for linoleum floor coverings of the 1920s and '30s. These timeless historic designs are applicable to any number of contemporary innovations. Designers, architects, and artists will be inspired both for interior applications and other art projects. This book is also an invaluable aid to anyone undertaking an authentic restoration of a 1920s-'30s-era domicile, the patterns herein will also inspire nostalgic infusions for today's art and architecture. The enclosed CD-Rom presents the images in a universal format, making it easy for designers to experiment with, manipulate, and apply images.

Library of Congress Control Number: 2005927602

Cover Design by Bruce Waters / Book Designed by Tina Skinner
Type set in Baveuse/Korinna BT

ISBN: 0-7643-2291-5
Printed in China

Published by Schiffer Publishing Ltd.
4880 Lower Valley Road, Atglen, PA 19310
Phone: (610) 593-1777; Fax: (610) 593-2002
E-mail: Info@schifferbooks.com

In Europe, Schiffer books are distributed byBushwood Books
6 Marksbury Ave., Kew Gardens, Surrey TW9 4JF England
Phone: 44 (0) 20 8392-8585; Fax: 44 (0) 20 8392-9876 E-mail: info@bushwoodbooks.co.uk
Free postage in the U.K., Europe; air mail at cost.

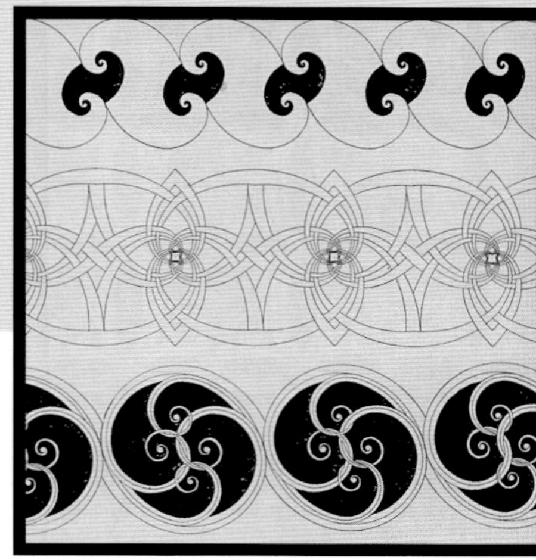

5

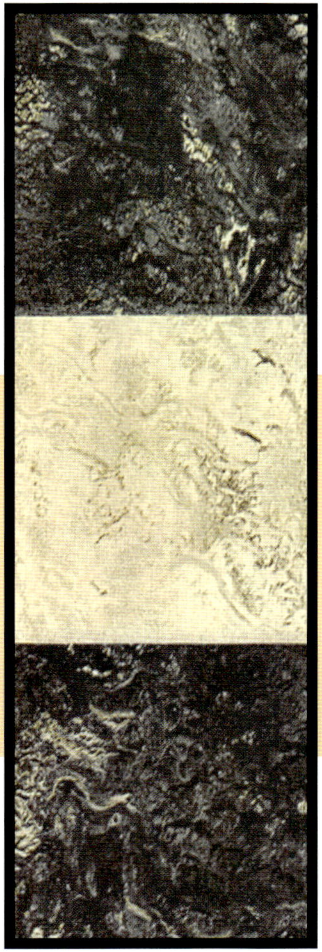

5

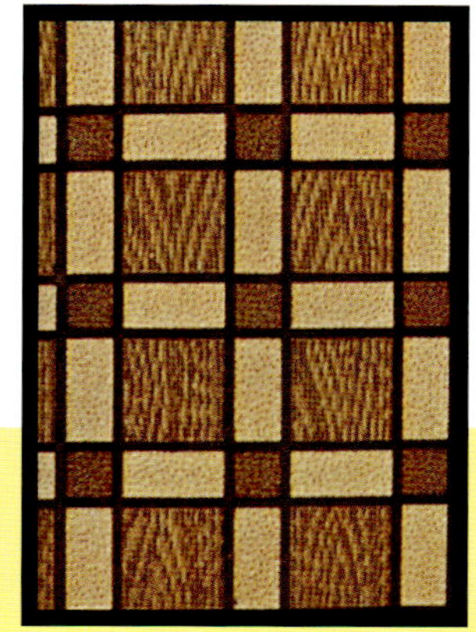

9

10

10

11

11

11

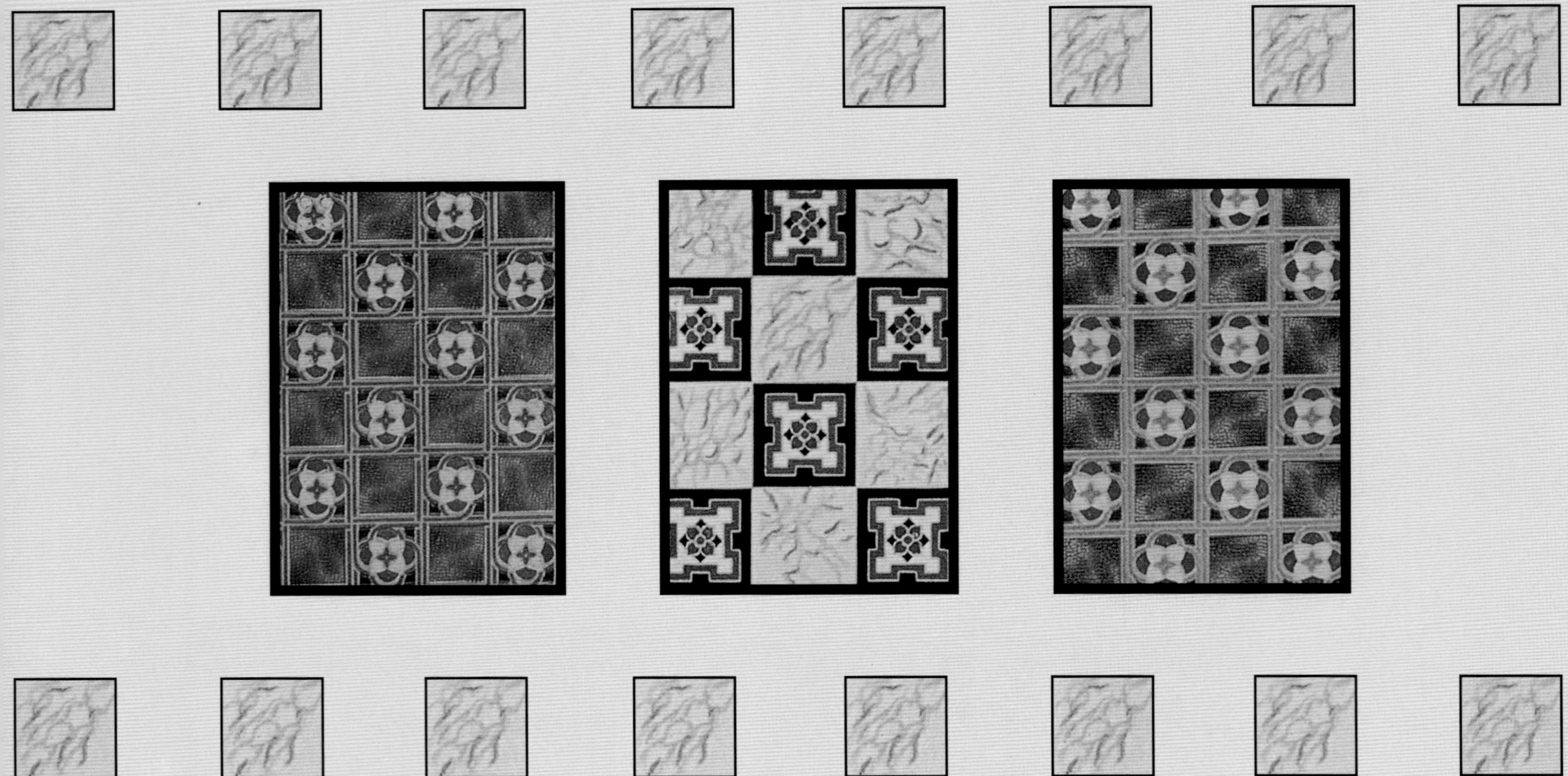

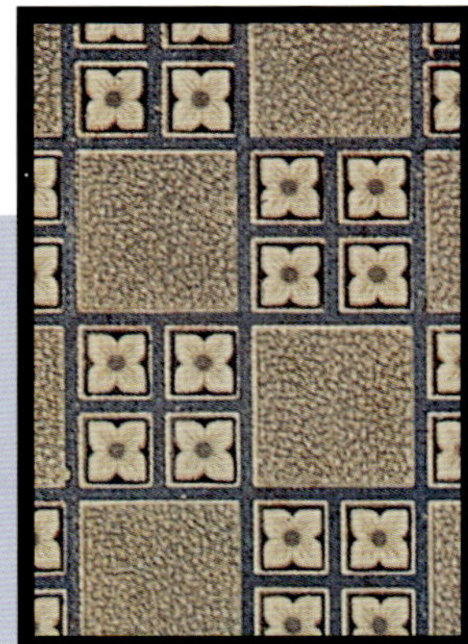

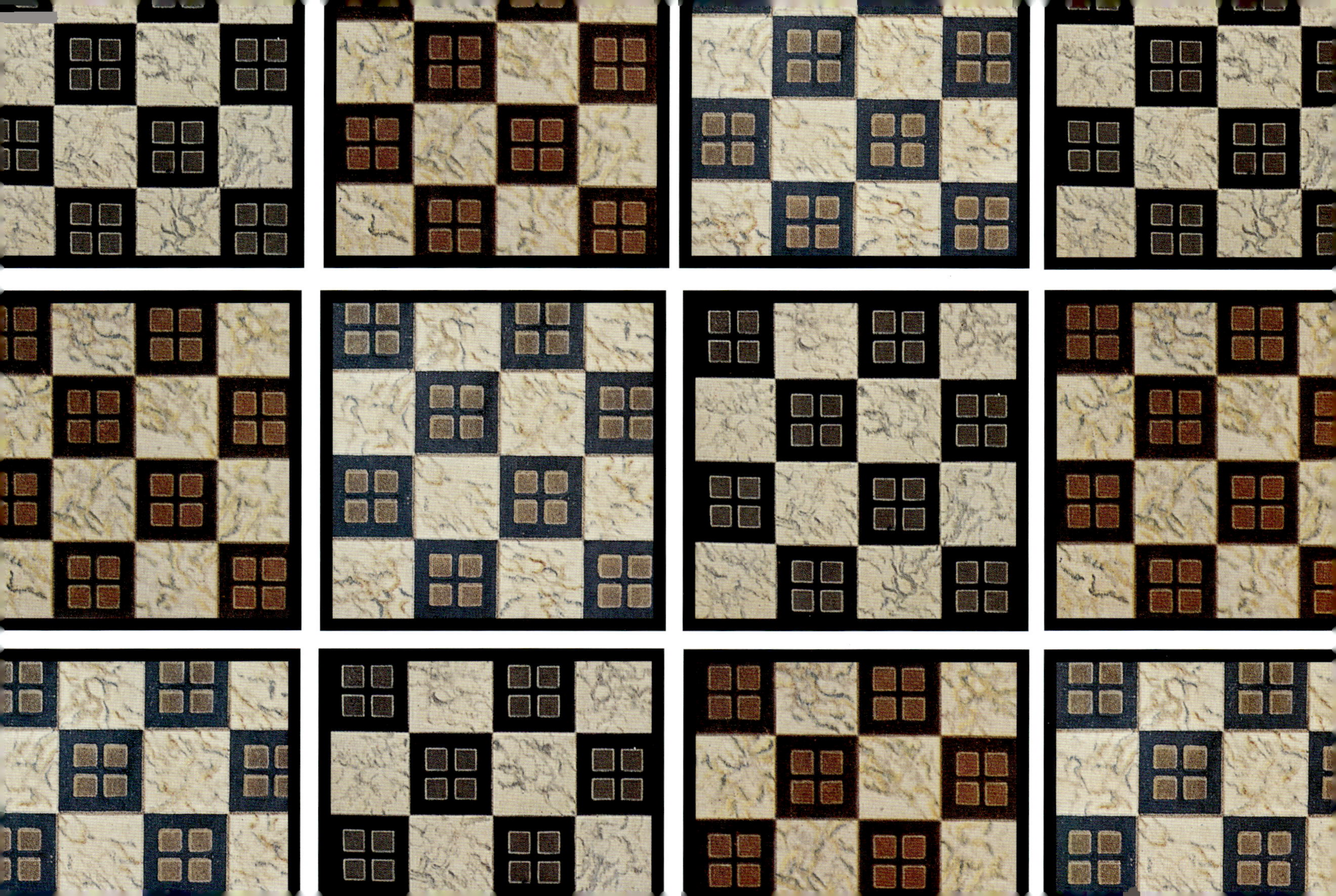

16

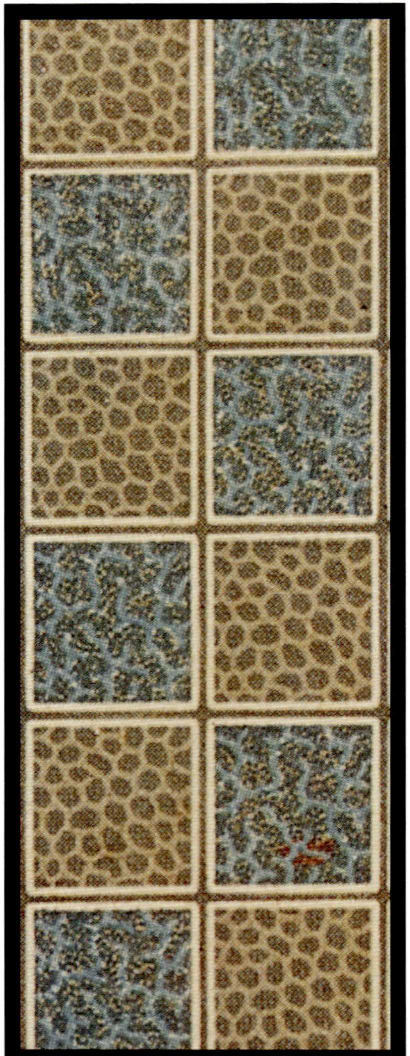

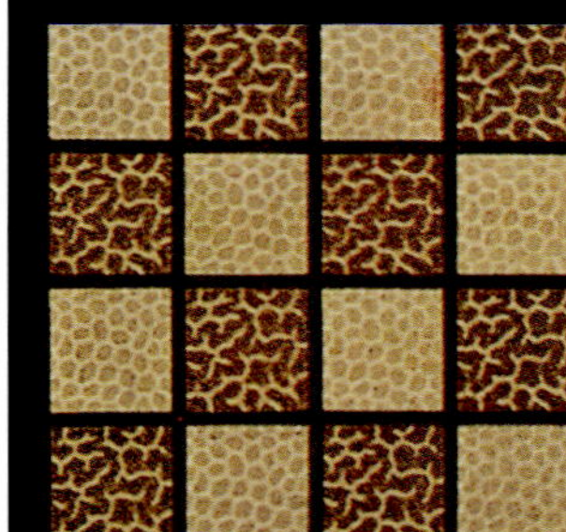

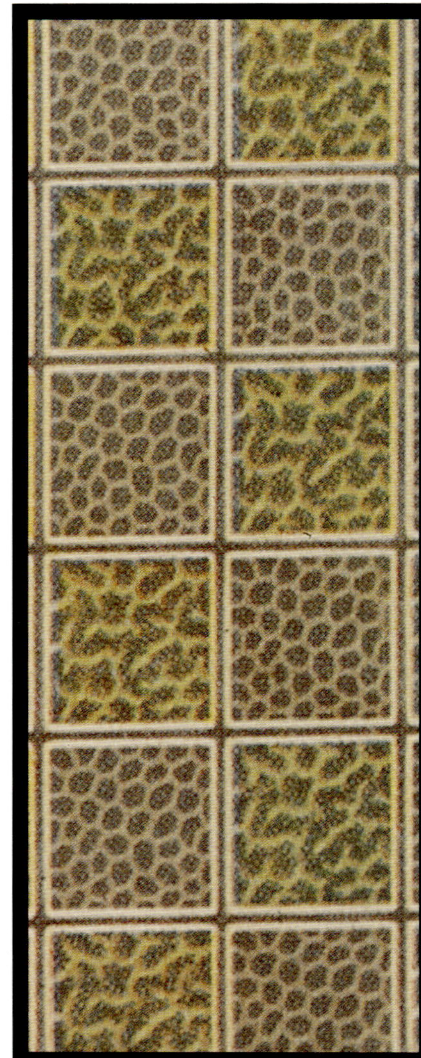

16

16

17

17

17

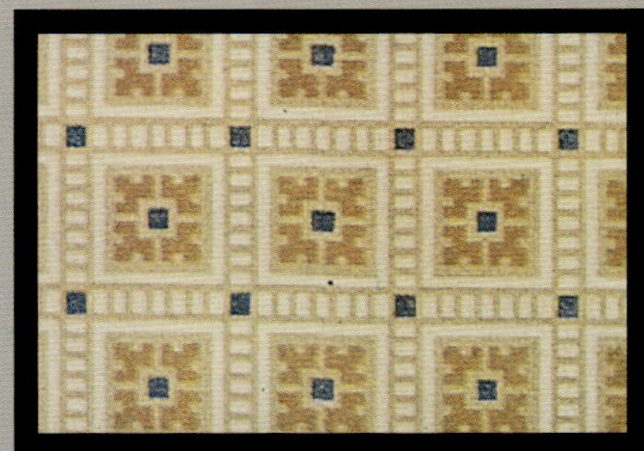

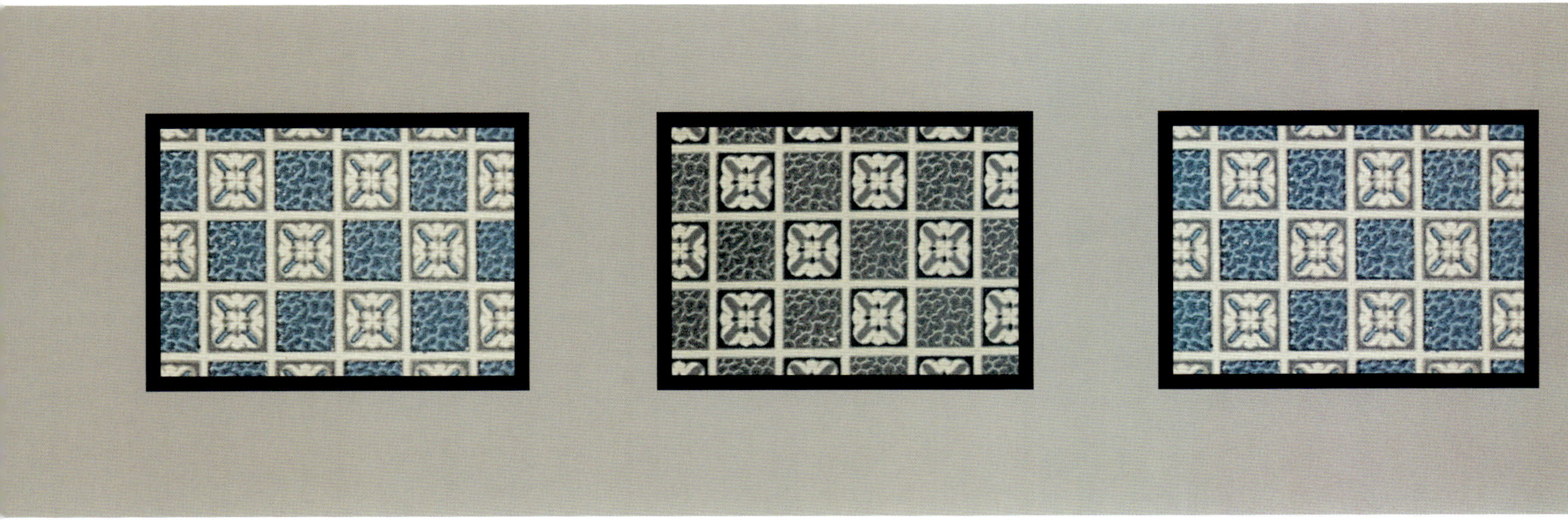

23

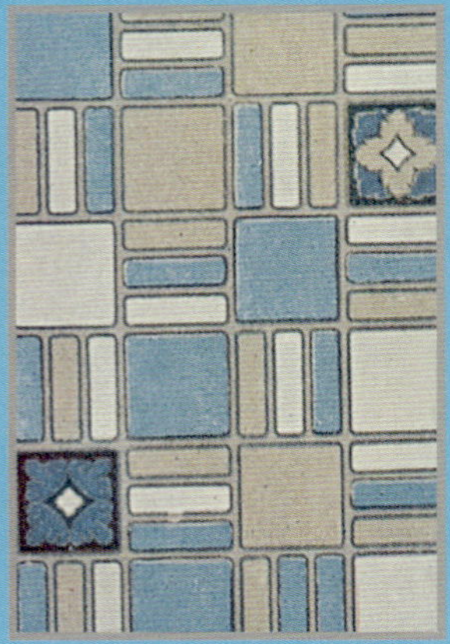

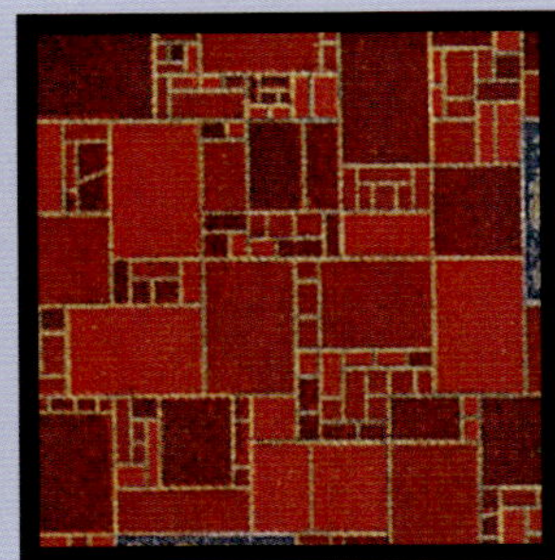

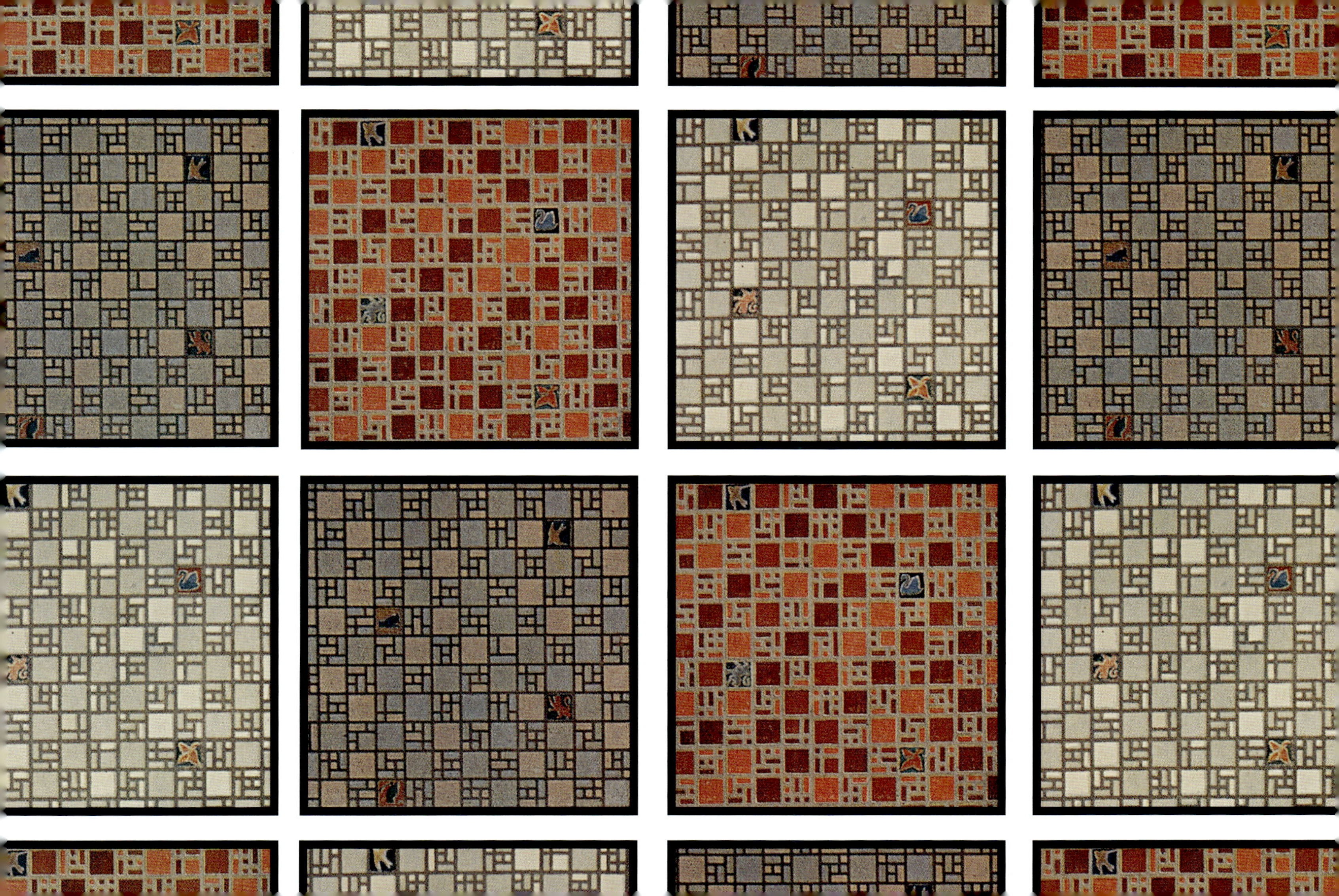

30

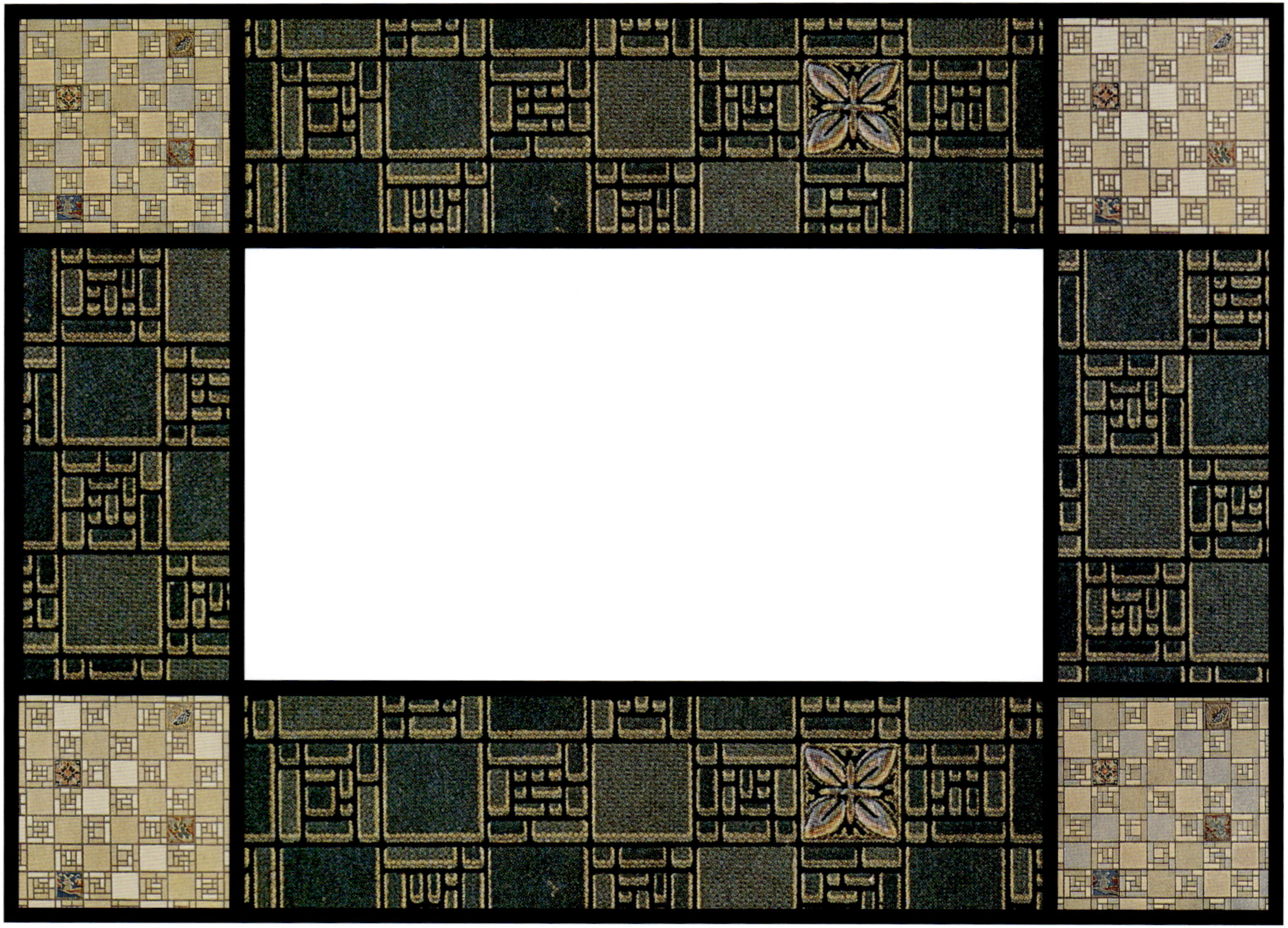

30

30

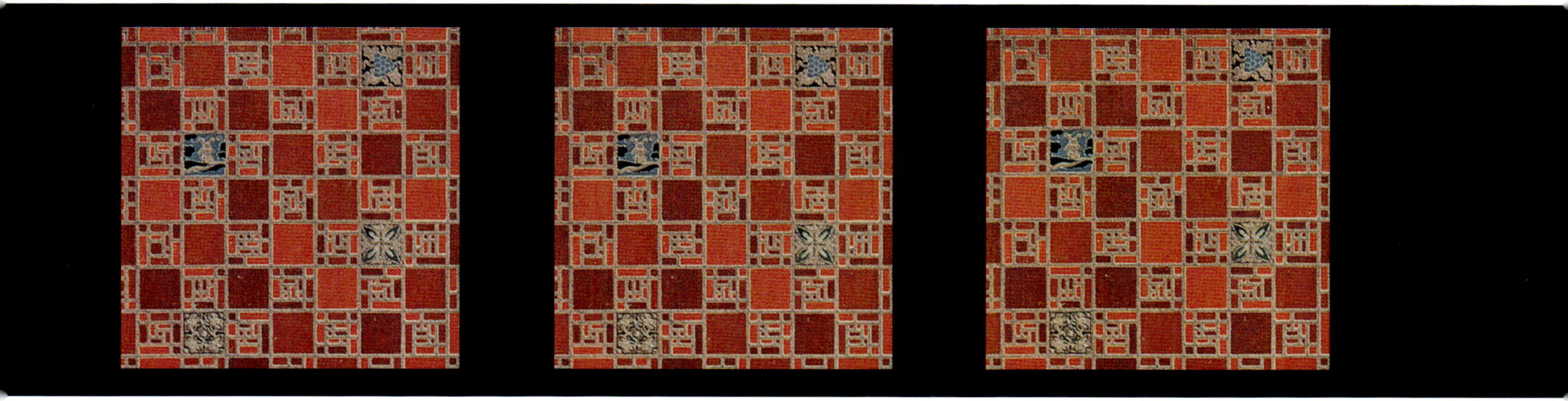

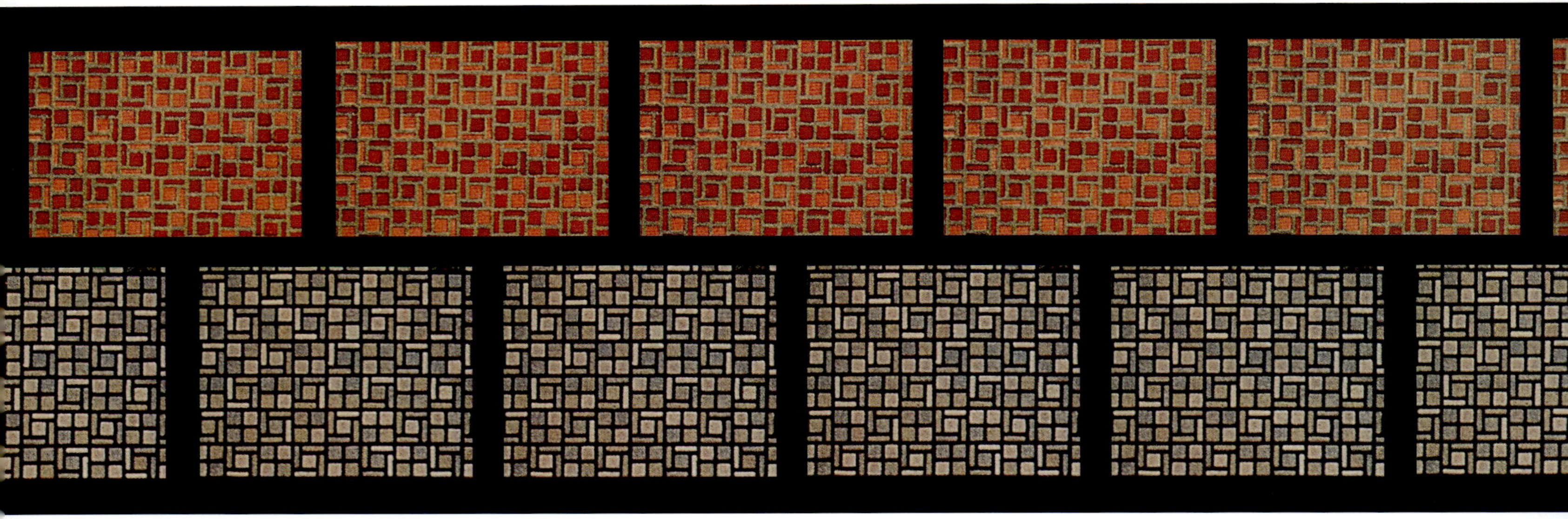

36

49

49

49

53

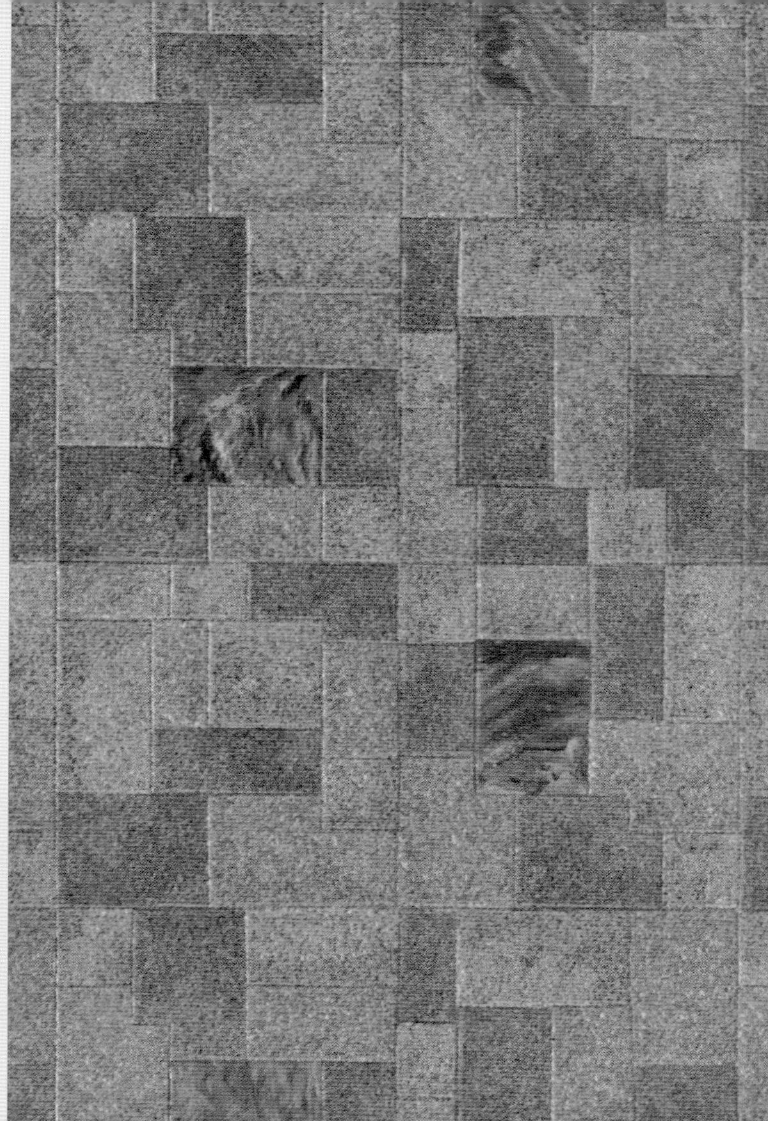

61

61

69

71

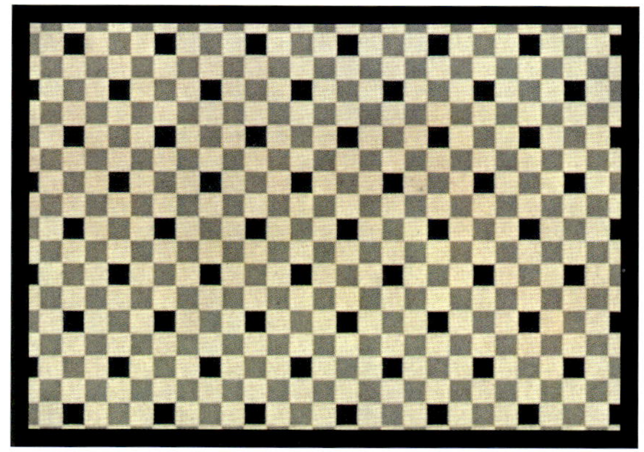

71

71

75

77

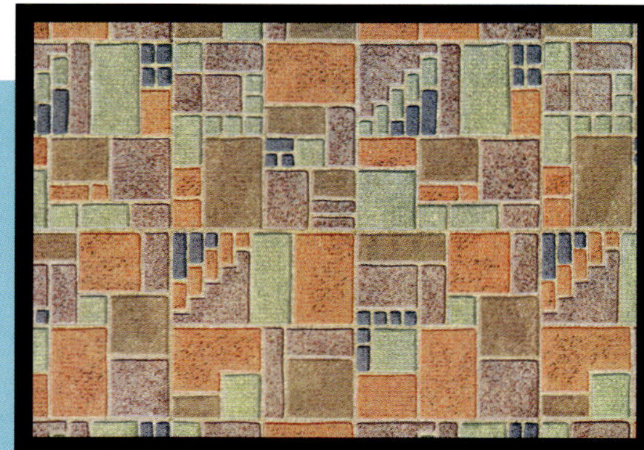

SCHIFFER PUBLISHING LTD
4880 LOWER VALLEY ROAD
ATGLEN, PA 19310-9717 USA

88

91

100

100

100

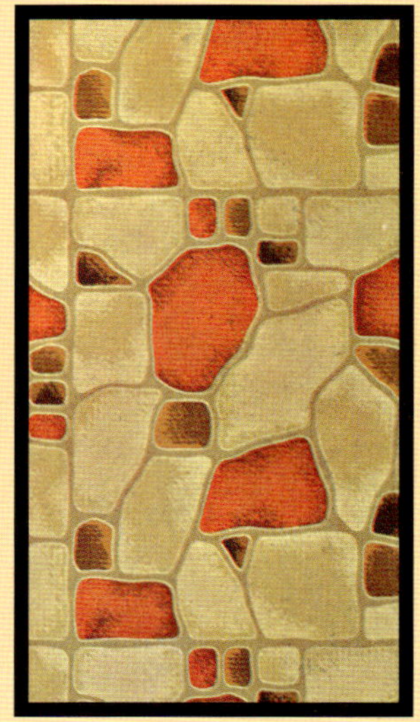

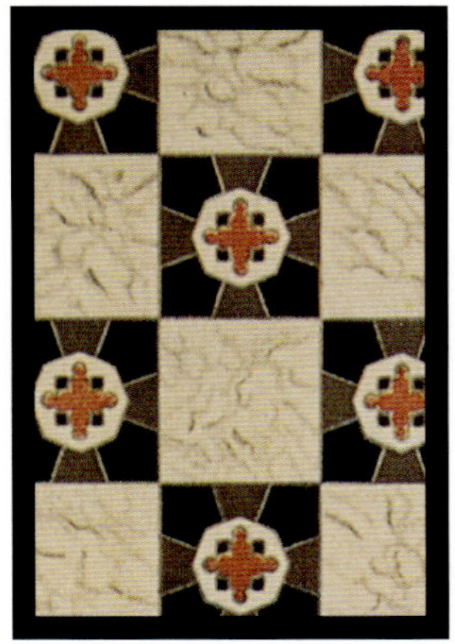

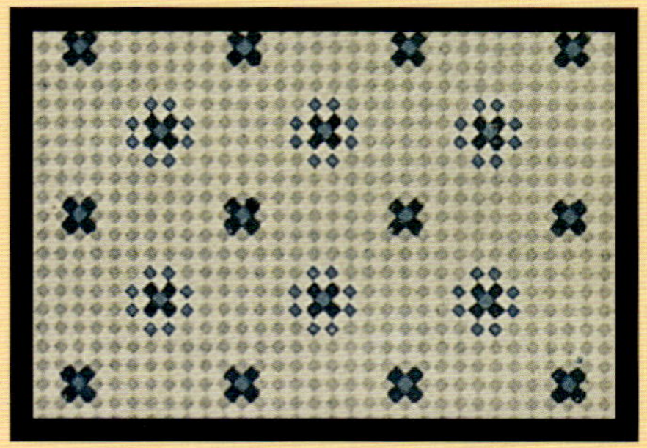

105

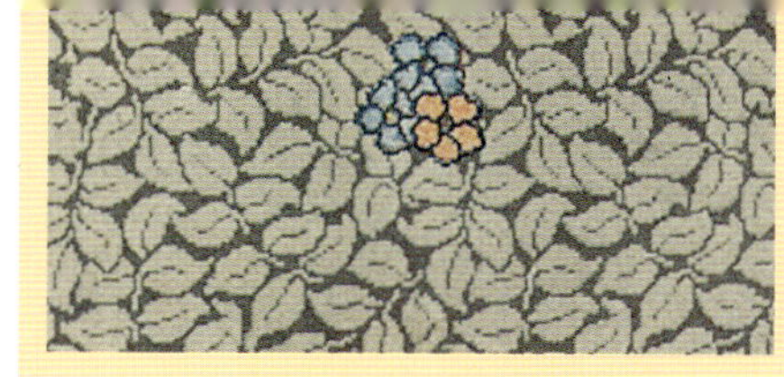

114

114

114

115

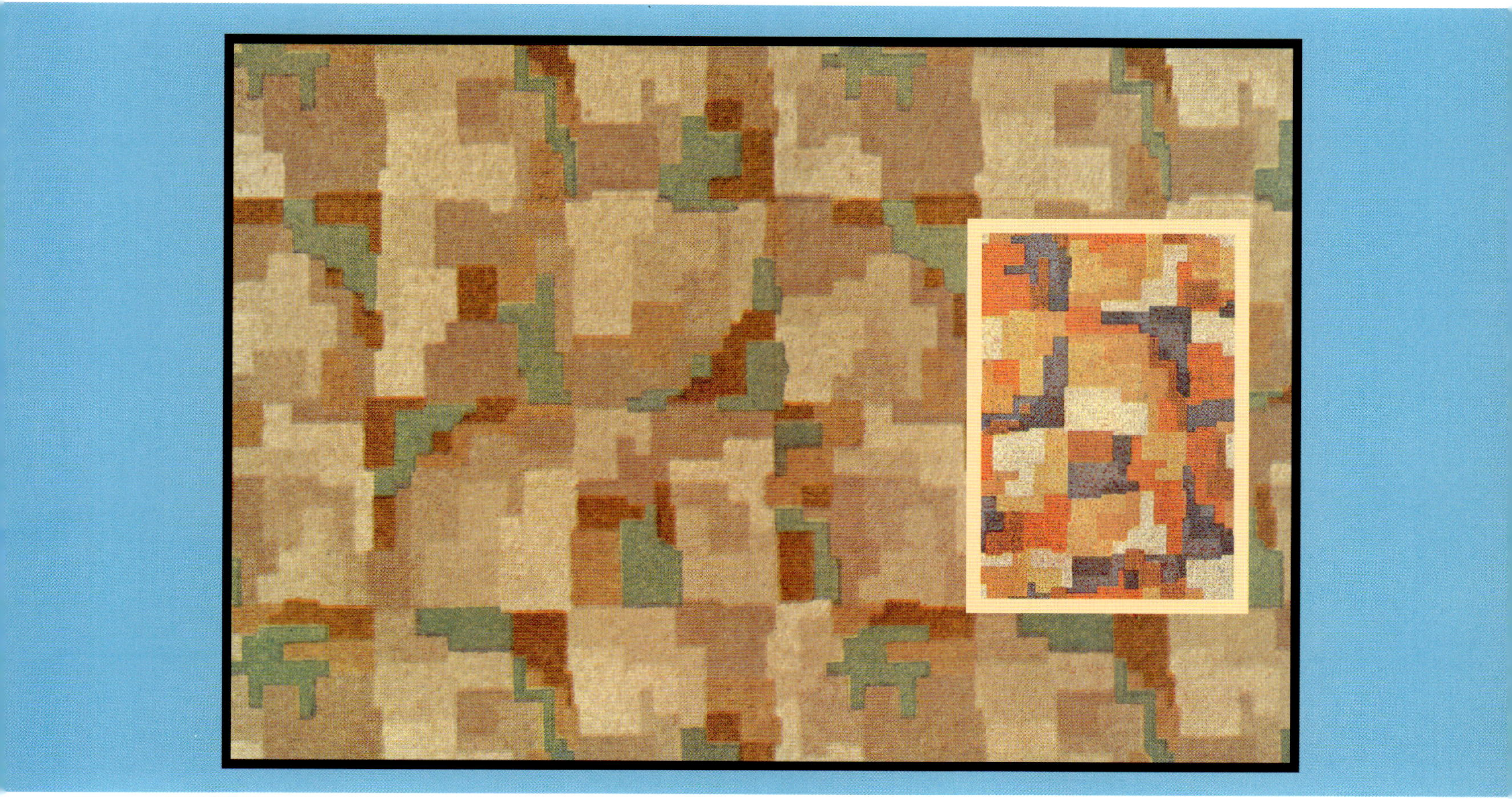

119

121

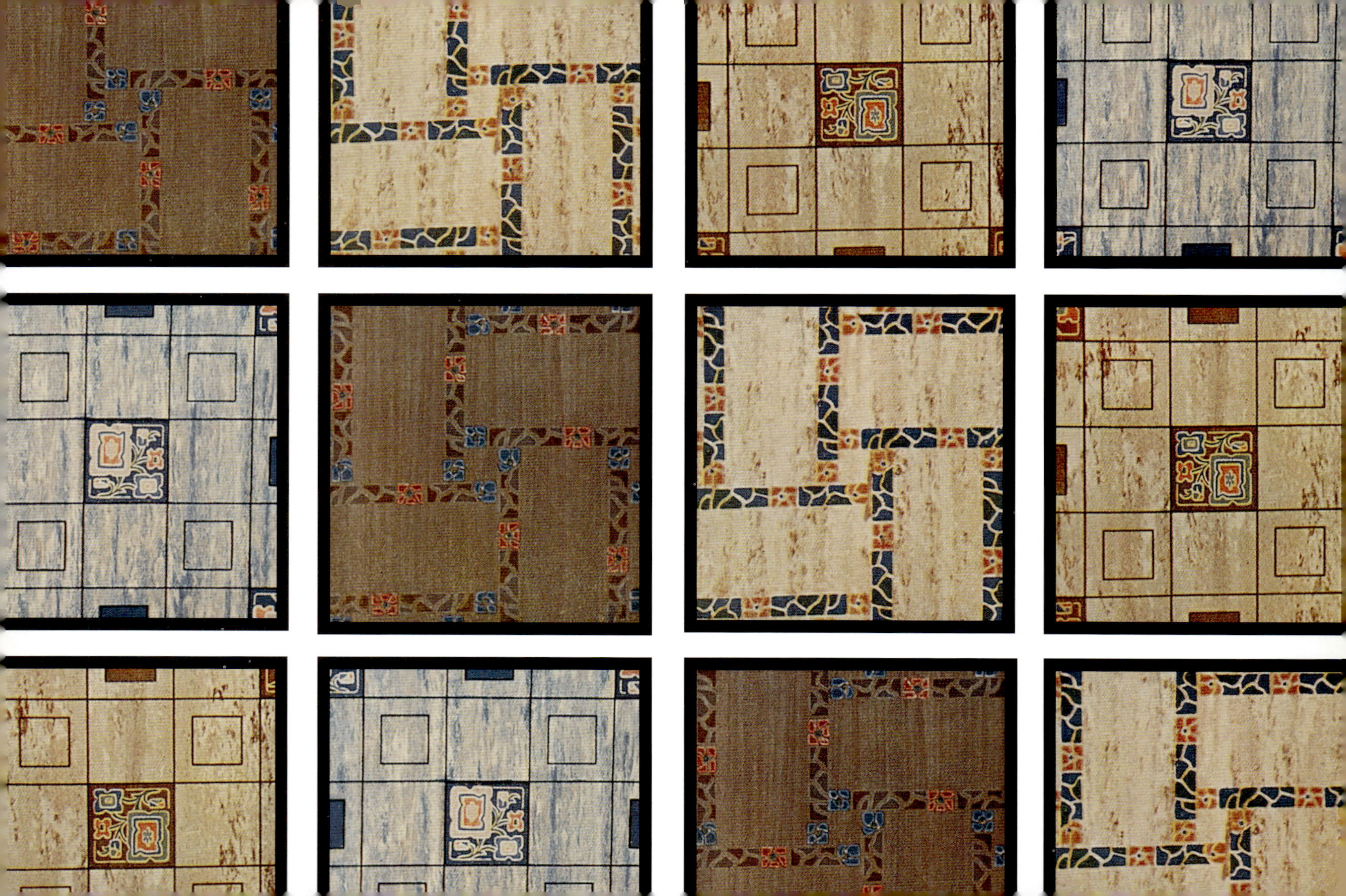

132

134

135

138

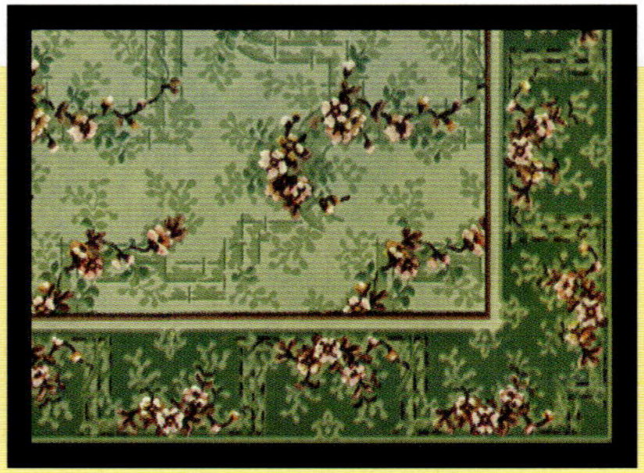

138

139

139

141

141

143

144

144

144

145

148

148

148

152

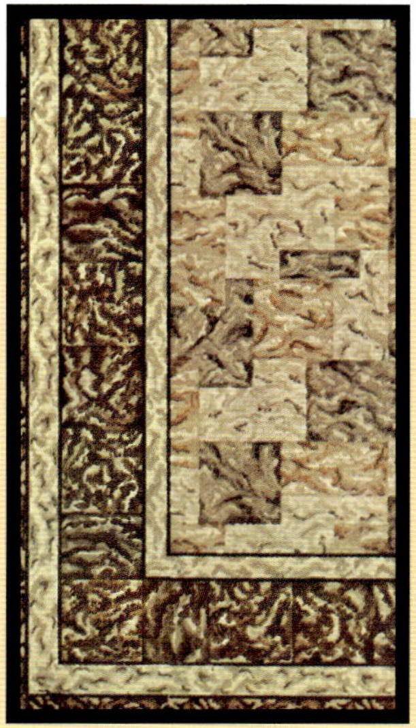

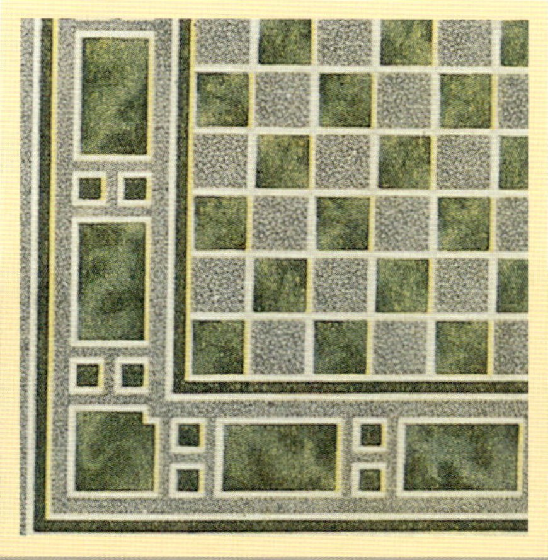

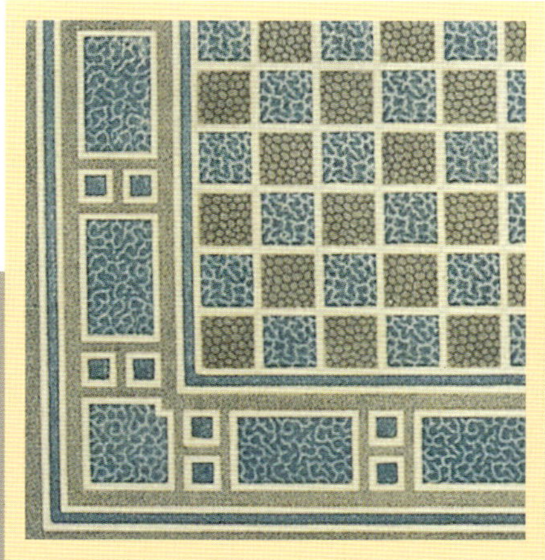

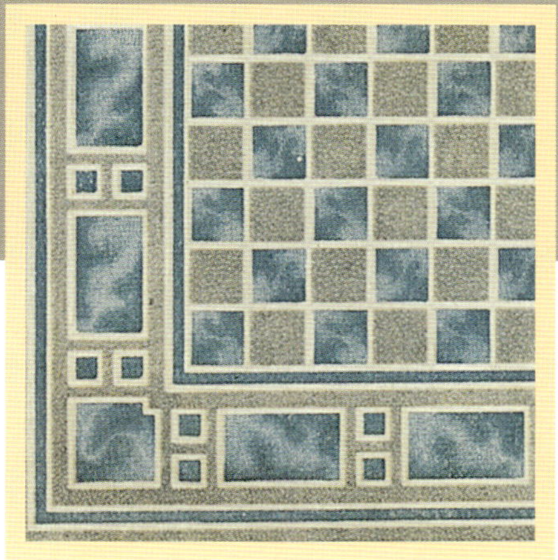

158

158

158

OTHER GREAT BOOKS FROM SCHIFFER PUBLISHING, LTD.

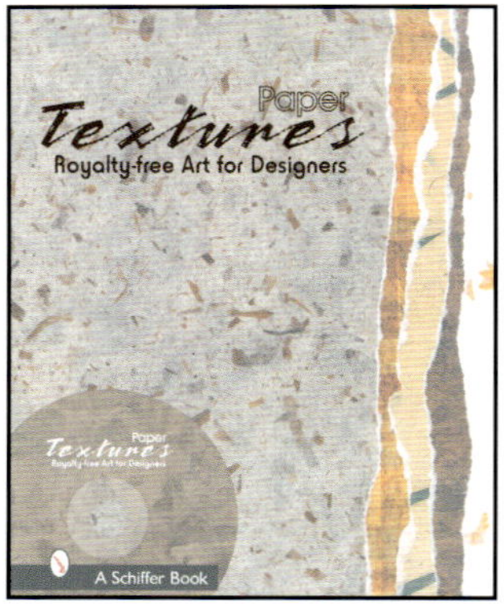

Paper Textures: Royalty Free Art for Designers. With two CDs.

Images of handmade papers, from soft and subtle to deeply marbled and highly textured. Examples of suggested Adobe® Photoshop applications are included along with the ready-to-use high resolution images.

Size: 8 1/2" x 11" * 114 color plates * 96 pp. * ISBN: 0-7643-2118-8 * Plastic cover with ring binding * $39.95

Historic Christmas Art: Santa, Angels, Poinsettia, Holly, Nativity, Children, and More Royalty-free Images on CD.

Size: 9" x 6" * 330 color illustrations * 176 pp.
ISBN: 0-7643-2120-X * hard cover * $24.95

Historic Holiday Art: New Year, Valentines, St. Patrick's Day, Easter, July 4th, Halloween, & Thanksgiving with 295 Royalty-free images on CD.

Size: 9" x 6" * 295 color illustrations * 176 pp.
ISBN: 0-7643-2119-6 * hard cover * $24.95

Art Nouveau Era Graphics: Ornamental Figures, Flowers, Emblems, Landscapes, and Animals with DVD.

A rich treasury of designs to inspire and enrich decorative arts projects. Originally presented in the 1890s for use in home interiors, ceramics, textiles, stationary, stained glass, and ironwork. In both print and digital form.

Size: 9" x 12" * 500 color and 275 b/w graphics * 152 pp. * ISBN: 0-7643-2042-4 * soft cover * $34.95